ODD ADAPTATIONS

WHY DO PITCHER PLANTS EAT BUGS?

AND OTHER ODD PLANT ADAPTATIONS

BY BRIANNA BATTISTA

Gareth Stevens

Please visit our website, www.garethstevens.com. For a free color catalog of all our high-quality books, call toll free 1-800-542-2595 or fax 1-877-542-2596.

Library of Congress Cataloging-in-Publication Data

Names: Battista, Brianna, author.
Title: Why do pitcher plants eat bugs? : and other odd plant adaptations / Brianna Battista.
Description: New York : Gareth Stevens Publishing, [2019] | Series: Odd adaptations | Includes index.
Identifiers: LCCN 2018003836| ISBN 9781538220313 (library bound) | ISBN 9781538220337 (pbk.) | ISBN 9781538220344 (6 pack)
Subjects: LCSH: Plants–Adaptation–Juvenile literature.
Classification: LCC QK912 .B38 2018 | DDC 581.4–dc23
LC record available at https://lccn.loc.gov/2018003836

First Edition

Published in 2019 by
Gareth Stevens Publishing
111 East 14th Street, Suite 349
New York, NY 10003

Designer: Sarah Liddell
Editor: Therese Shea

Photo credits: Cover, p. 1 (pitcher plant) Skynavin/Shutterstock.com; background used throughout Captblack76/Shutterstock.com; cover, pp. 1 (fly), 25 (bottom) irin-k/Shutterstock.com; p. 4 orangecrush/Shutterstock.com; p. 5 Naruedom Yaempongsa/Shutterstock.com; p. 6 Mark Purches/Shutterstock.com; p. 7 kram9/Shutterstock.com; pp. 8, 18 KarenHBlack/Shutterstock.com; p. 9 (sun) ESB Professional/Shutterstock.com; p. 9 (plant) Ilya Andriyanov/Shutterstock.com; p. 9 (boy) michaeljung/Shutterstock.com; p. 10 Bernadette Heath/Shutterstock.com; p. 11 akphotoc/Shutterstock.com; p. 12 (top and bottom) Petr Salinger/Shutterstock.com; p. 13 (top) Steve Brigman/Shutterstock.com; p. 13 (bottom) Wolfgang Kaehler/Contributor/LightRocket/Getty Images; p. 14 allstars/Shutterstock.com; p. 15 Wanida_Sri/Shutterstock.com; p. 16 yelantsevv/Shutterstock.com; p. 17 Byelikova Oksana/Shutterstock.com; p. 19 Rawdon Sthradher/Shutterstock.com; p. 20 Natalia Ramirez Roman/Shutterstock.com; p. 21 DEA/C. DANI/I. JESKE/Contributor/De Agostini; p. 22 Lee Prince/Shutterstock.com; p. 23 David Tipling/Lonely Planet Images/Getty Images; p. 24 PicTrans/Wikimedia Commons; p. 25 (top) Sergei Drozd/Shutterstock.com; p. 25 (right) Fotokostic/Shutterstock.com; p. 25 (left) Alekcey/Shutterstock.com; p. 26 Mohd KhairilX/Shutterstock.com; p. 27 Anna ART/Shutterstock.com; p. 28 Dick van Toom/Wikimedia Commons; p. 29 TED MEAD/Photolibrary/Getty Images.

Printed in the United States of America

CPSIA compliance information: Batch #CS18GS: For further information contact Gareth Stevens, New York, New York at 1-800-542-2595.

CONTENTS

Plant Power 4

Botany 101 8

Thirsty Plants 10

Desert Tricks 12

Too Much H_2O 14

Burn, Baby, Burn. 16

Meat-Eating Plants! 20

Parasitic Plants. 22

Amazing Mimics. 24

Can't Touch This! 28

Glossary. 30

For More Information .31

Index 32

Words in the glossary appear in **bold** type the first time they are used in the text.

PLANT POWER

Plants are amazing! They give us food to eat and produce oxygen we need to breathe. From the tallest tree to the tiniest moss, there are at least 400,000 different species, or kinds, of plants.

Have you ever thought about why some plants act and look the way they do? Some vegetables grow on tall vines, and others grow underground. Some flower stems feel smooth, and others are prickly. These different features are called adaptations. An adaptation is the way a species changes how it looks or behaves over time, based on what it must do to survive in its **environment**.

TOMATOES ON A VINE

THANK YOU, PLANTS!

The study of plants is called botany. Botany can teach us a lot about the world. It's fun, too! If land plants hadn't **developed** more than 450 million years ago, Earth's air would hold too much carbon dioxide gas for humans—and most creatures—to live!

CAN YOU IMAGINE A WORLD WITHOUT PLANTS? PLANTS GIVE US AIR TO BREATHE, FOOD TO EAT, SHADE TO ENJOY, AND SO MUCH MORE!

Adaptations can range from the expected to the unbelievably odd. At the heart of all adaptations is the need for the organism to survive. Plants can't run, swim, or hide from animals. For example, if a dog decides to dig in a spot covered with grass and wildflowers, these plants are out of luck. They'll probably be uprooted and die.

DO OR DIE

Nineteenth-century scientist Charles Darwin argued that all organisms are competing to survive, and those that adapt better are more likely to reproduce and pass those adaptations on. Organisms that don't adapt as well have fewer young and finally die out. Darwin described his ideas in the 1859 book *On the Origin of Species by Means of Natural Selection*.

OVER TIME, SOME PLANT BODIES AND BEHAVIORS HAVE ALTERED TO PROTECT THE PLANTS AND EVEN HELP THEM FLOURISH. It's useful to understand the basics of how most plants function in the world before exploring the special ways they've adapted to their environments.

SCIENTISTS THINK THE FIRST PLANTLIKE LIFE-FORMS CAME TO EXIST AFTER AN ORGANISM CALLED AN ALGA SWALLOWED A TYPE OF BACTERIA ABOUT 1.6 BILLION YEARS AGO.

BOTANY 101

Have you ever heard that talking to your plants is good for them? It's actually true! When humans breathe out, we **release** carbon dioxide. **THE MAJORITY OF PLANTS NEED CARBON DIOXIDE TO MAKE THEIR OWN FOOD THROUGH THE PROCESS OF PHOTOSYNTHESIS.** They also need sunlight, water, and chlorophyll, which is green matter found in most plants.

Plants usually have roots, stems, leaves, and special **tissues** that carry water and food throughout the plant. Most also need soil to grow in. The soil gives the plants water and **nutrients**. There are plants called epiphytes that grow on taller plants, however. They don't need soil!

EPIPHYTE

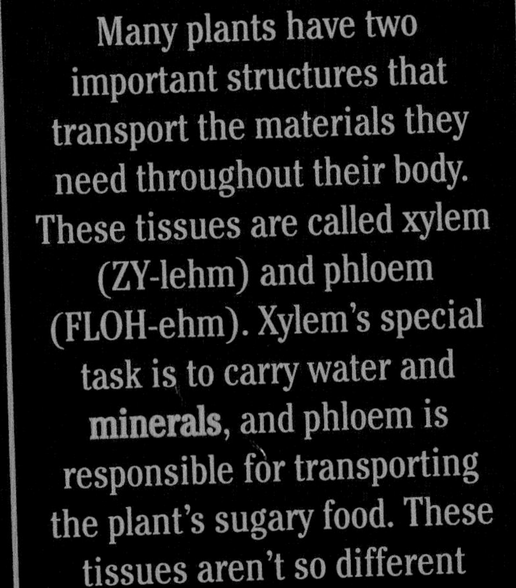

TERRIFIC TISSUES

Many plants have two important structures that transport the materials they need throughout their body. These tissues are called xylem (ZY-lehm) and phloem (FLOH-ehm). Xylem's special task is to carry water and **minerals**, and phloem is responsible for transporting the plant's sugary food. These tissues aren't so different from veins in our bodies!

WHAT HAPPENS IN PHOTOSYNTHESIS?

THIRSTY PLANTS

All plants need water to grow. How do they survive in environments where it's very hot and doesn't rain much? The desert is one such challenging place to live, but plants there have found ways to thrive.

When you think of plant life in the desert, a cactus may come to mind! There are about 2,000 species of these plants. Cacti don't usually have leaves because leaves release water into the air. Cacti can store water in their tissues in order to survive long periods of **drought**.

THE THICK, WAXY SKIN ON A CACTUS'S STEM IS A STRONG WALL THAT HOLDS WATER IN.

ROOT ADAPTATIONS

Cacti roots don't go into the soil very deep. **INSTEAD OF GOING FAR DOWN INTO THE GROUND, THE ROOTS STAY CLOSE TO THE SURFACE AND SPREAD OUT OVER A WIDE AREA.** During a rain shower, the roots are able to catch as much rainfall as possible!

MOST CACTI ARE COVERED IN SHARP SPINES, OR NEEDLES. THIS IS AN ADAPTATION, TOO! THE SPINES HELP PROTECT THE CACTUS FROM HUNGRY OR THIRSTY ANIMALS.

DESERT TRICKS

Cacti are cool, but you can find even stranger adaptations in the desert. **THE DESERT RHUBARB HAS A WAY TO WATER ITSELF!** This plant, which grows in the Middle East, has broad leaves with channels that funnel rainwater to the plant's roots.

A fern known as the false rose of Jericho learned to "play dead." When the weather gets very dry, the plant turns brown and rolls its stems into a tight ball. **IT LOOKS DEAD, BUT IT CAN SURVIVE FOR YEARS LIKE THIS!** When it finally rains, the plant returns to its bright green fern form.

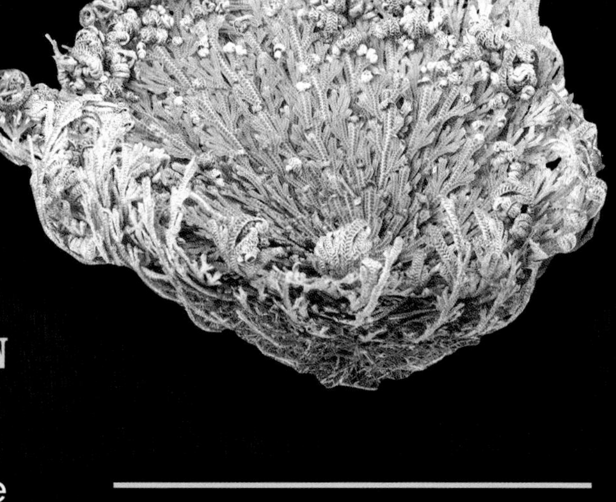

THE FALSE ROSE OF JERICHO CAN EVEN BE UPROOTED AND BLOWN AROUND. WHEN THE RAIN FINALLY COMES, THE PLANT RETURNS TO LIFE!

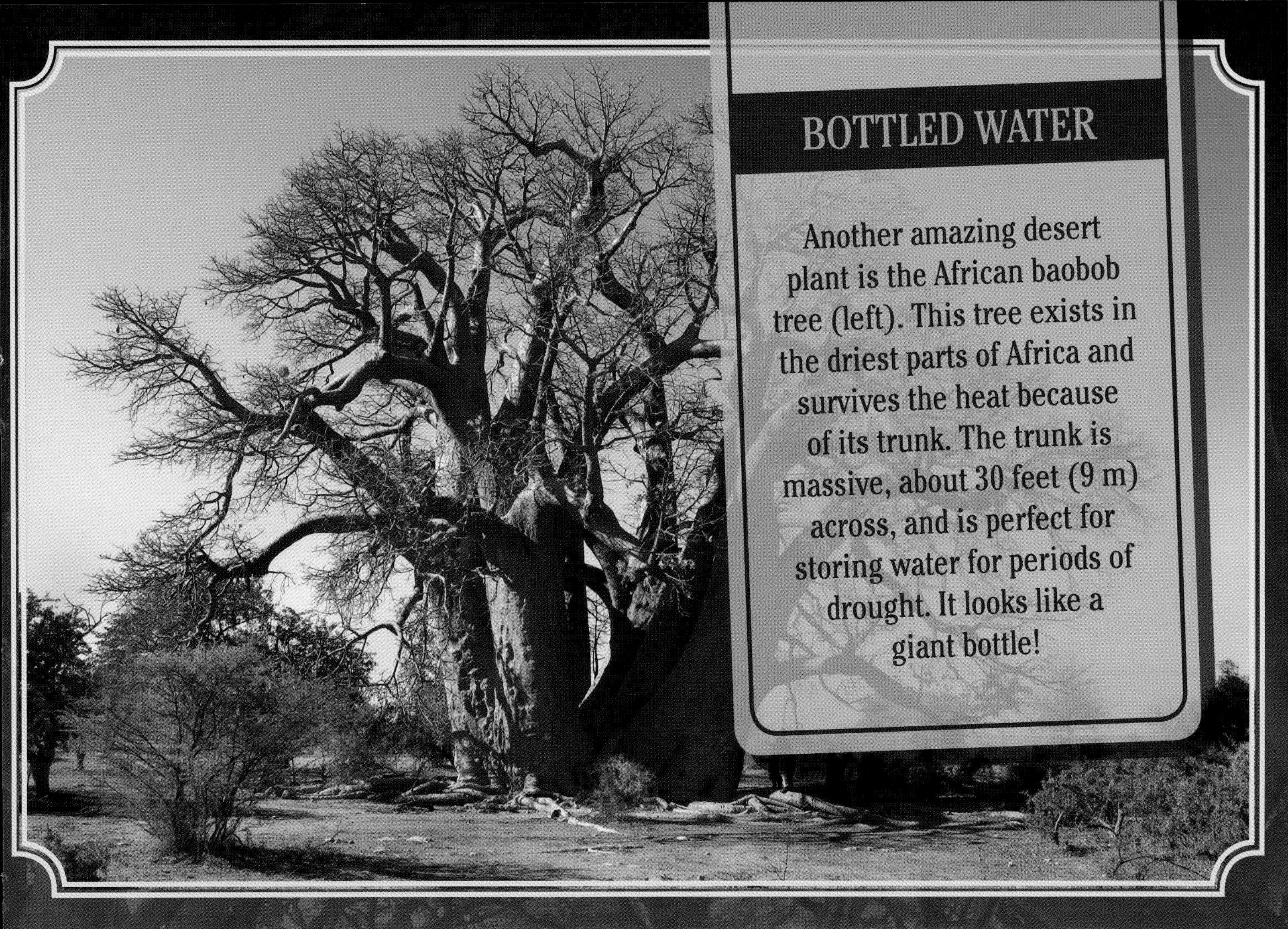

BOTTLED WATER

Another amazing desert plant is the African baobob tree (left). This tree exists in the driest parts of Africa and survives the heat because of its trunk. The trunk is massive, about 30 feet (9 m) across, and is perfect for storing water for periods of drought. It looks like a giant bottle!

DESERT RHUBARB

TOO MUCH H_2O

Can plants ever have too much water? The answer is yes! Plants need water to be able to produce food, but too much water can attract, or draw, harmful bacteria and **fungi**. Rain forest plants have found several different ways to adapt.

Some rain forest plants have leaves shaped like spouts. Extra water drips out of these spouts as if it's coming out of a teapot! Other rain forest plants have surfaces that absorb, or take in, water more slowly. **RAIN FOREST PLANTS MAY EVEN CATCH RAINFALL AND HOLD IT UNTIL IT'S NEEDED.** Some have leaves that are resistant to growing harmful fungi, too.

PLANT ARCHITECTURE

Sometimes soil has too much water in it for plant growth. The soil along oceans and rivers can be soft and muddy. Trees called mangroves adapted to these conditions by spreading their roots out wide above the soil. The roots support the main trunk above the ground. **THEY TRAP NUTRIENTS, TOO!**

THE MANGROVE'S SPECIAL ABOVEGROUND "BREATHING ROOTS" ARE ABLE TO TAKE IN OXYGEN THE PLANT NEEDS, WHICH IS OFTEN LACKING IN WETLAND SOIL.

BURN, BABY, BURN

Fire is one of the world's most destructive natural forces. Wildfires occur in forests, grasslands, and other environments all over the planet. Fire can kill people and consume massive buildings, yet some plants have adapted to withstand it!

The giant sequoia tree—which can grow to be over 200 feet (61 m) tall—has thick bark protecting its important tissues. Its fire-resistant bark can be up to 2 feet (61 cm) thick! Wildfires can harm these trees, but they usually survive. **GIANT SEQUOIAS EVEN BENEFIT FROM WILDFIRES. THE ASH THAT REMAINS IN THE SOIL AFTER A FIRE ACTS AS FERTILIZER!**

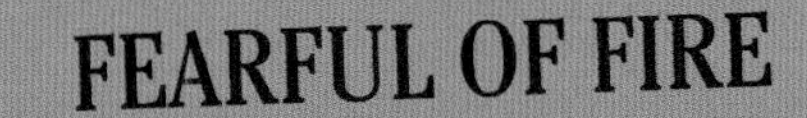

FEARFUL OF FIRE

In 2013, a fire broke out in Yosemite National Park in California's Sierra Nevada. The public was concerned that the fire would destroy the giant sequoias. The park service knew that the giant sequoias would be fine, but to make people feel better, they set up sprinklers near the trees!

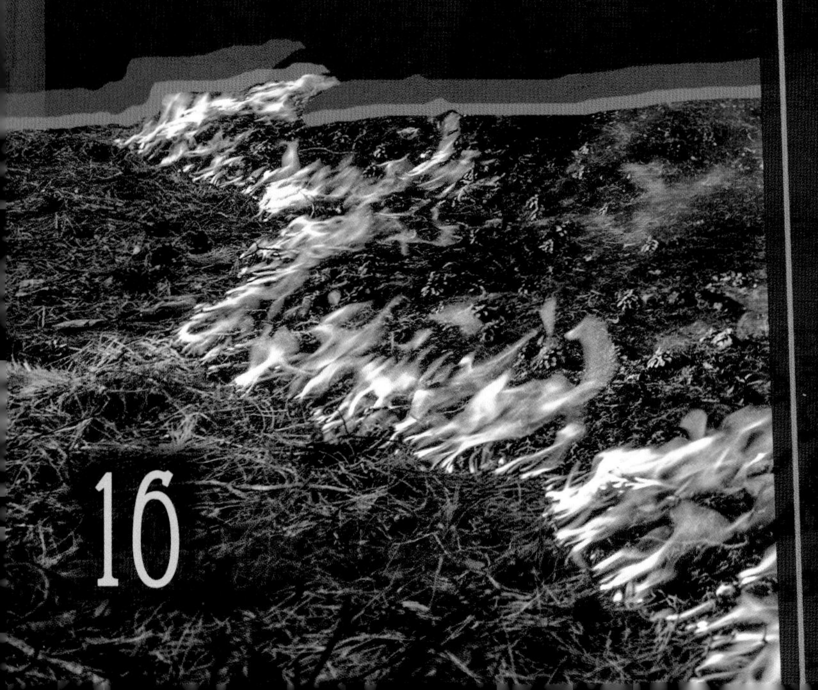

MOST GIANT SEQUOIAS ARE FOUND IN CENTRAL CALIFORNIA. PEOPLE TRAVEL FROM ALL OVER THE WORLD TO VISIT THESE MASSIVE TREES WITH BEAUTIFUL RED BARK.

SOME PLANTS HAVE SEEDS THAT CANNOT BE RELEASED WITHOUT FIRE! Many species of shrubs and trees called banksia live in areas of Australia where fires are common. Banksias' seeds are in woody structures grouped together in the shape of a cone. For most species of this plant, the seed **follicles** open only in extreme heat.

This adaptation is helpful for more than just withstanding fire. The remains of the fire also help the fallen seeds germinate, or begin to grow. Much of the banksias' competition for soil and nutrients has burned away in the wildfire, and the soil becomes fertilized with ash. It's perfect for the growth of banksia seeds.

THIS IMAGE SHOWS THE "CONES" OF A BANKSIA TREE AFTER A FIRE. SOON, THE SEEDS WILL BE RELEASED!

SMART BLOOMS

The Australian grass tree is another plant that benefits from fire. Extreme heat causes it to bloom. This is an advantage for the tree because there aren't as many flowers in bloom after a fire. The grass tree's tall blooms have less competition for **pollinating** insects!

MEAT-EATING PLANTS!

The pitcher plant lives in places with soil that lacks nutrients. However, it has adapted to get nutrients a different way. **THE PITCHER PLANT IS A CARNIVOROUS PLANT, WHICH MEANS IT EATS MEAT!** Carnivorous plants don't eat hamburgers or steak, though. They trap and eat insects.

There are several kinds of pitcher plants. They may look different, but all have leaves shaped to capture bugs. The plants produce nectar, a sweet liquid that attracts insects. Insects crawl into the tube seeking nectar, but fall into **digestive** juices at the bottom of the "pitcher." **PITCHER PLANTS ABSORB NUTRIENTS FROM DIGESTED INSECTS THROUGH THEIR LEAVES!**

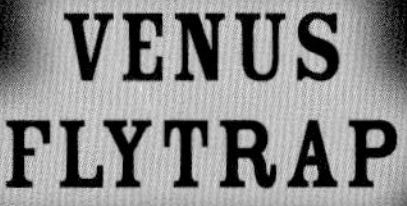

FEED ME!

The most famous carnivorous plant is the Venus flytrap. This organism has two leaves that look like jaws. The plant attracts prey with nectar within its trap. Its leaves shut when tiny hairs sense pressure from an insect. **WHEN THE INSECT WALKS ACROSS THE HAIRS, THE TRAP SHUTS IN ABOUT A HALF A SECOND!**

NEPENTHES IS A PITCHER PLANT THAT'S ALSO CALLED A MONKEY CUP. SMALL MONKEYS SOMETIMES DRINK FROM THE WATER THAT COLLECTS IN IT!

PARASITIC PLANTS

Did you know that plants can be parasites? Parasites live on or in other organisms, harming the organisms they benefit from. **PARASITIC PLANTS ATTACH THEMSELVES TO OTHER PLANTS, CALLED HOSTS.** Sometimes they steal water or nutrients from their hosts. Host plants often can't make enough food to support themselves and the parasites. The hosts may die as a result.

Other parasitic plants use hosts as supports. Strangler figs, for example, may begin life as a sticky seed left by an animal high up a tree. Their roots grow down, sometimes surrounding a tree's trunk, until they hit soil. The host tree may die from lack of sunlight and nutrients.

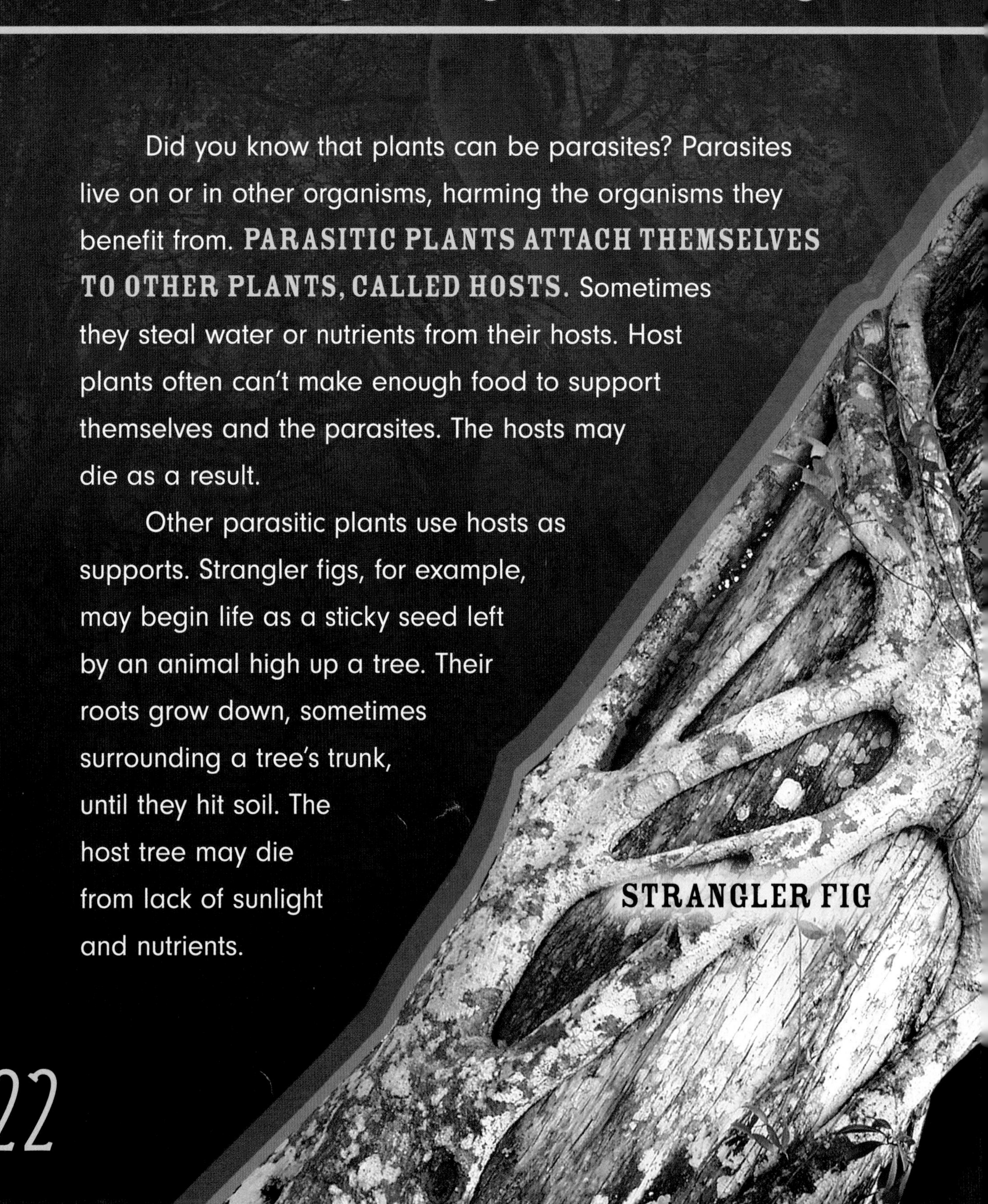

STRANGLER FIG

MENACING MISTLETOE

Mistletoe is known for being the holiday plant that people kiss under, but it's not quite so sweet in nature. **MISTLETOE IS A PARASITIC VINE THAT ATTACHES TO TREES AND STEALS THEIR WATER AND NUTRIENTS.** The plant is able to photosynthesize food on its own, but it also takes food from other plants.

AMAZING MIMICS

Some plants have learned to mimic other organisms! Mimicry is an adaptation that makes one organism look like another for a purpose. Mimicry helps plants trick, attract, or hide from insects and other animals.

The Australian bird orchid is a beautiful flowering plant that uses mimicry to attract pollinating insects. **ONE PART OF THE ORCHID LOOKS AND EVEN SMELLS JUST LIKE A FEMALE WASP!** Male wasps are drawn to the flower. When they swoop down hoping to find a mate, they get covered in pollen. The wasps spread it to the next orchid they visit, helping the flowers reproduce.

THIS BIRD ORCHID IS MIMICKING A WASP. WOULD YOU BE TRICKED?

STAY LOW

To further trick male wasps, the bird orchid has adapted a behavior, too. It grows low along the ground. That's because female wasps don't have wings and stay close to the ground when ready to mate. In fact, male wasps overlook orchids that have grown too tall!

HOW INSECTS POLLINATE FLOWERS

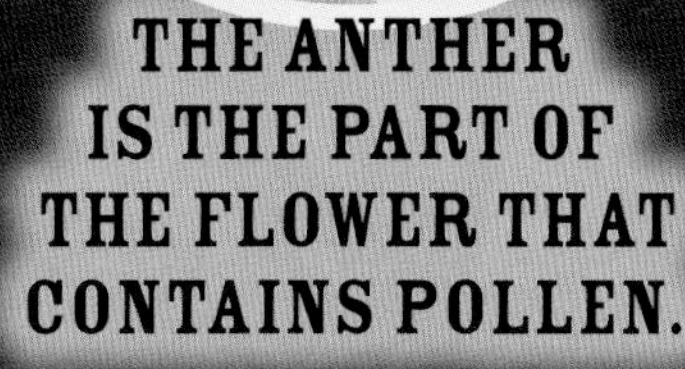

1

THE ANTHER IS THE PART OF THE FLOWER THAT CONTAINS POLLEN.

2

AN INSECT VISITS THE FLOWER FOR THE NECTAR AND PICKS UP STICKY POLLEN.

3

THE INSECT CARRIES THE POLLEN TO THE NEXT FLOWER. POLLEN FALLS ONTO THE PISTIL, WHICH IS THE PART OF A FLOWER THAT LEADS TO THE PLANT'S EGGS.

4

THE EGG CELLS ARE FERTILIZED BY POLLEN AND GROW INTO SEEDS!

One of the most fascinating mimicking adaptations is seen—or smelled—in the plant *Rafflesia arnoldii.* A common name for this plant is the corpse lily. **IT SMELLS LIKE ROTTING MEAT!** The flower smells bad in order to attract bugs, called carrion flies, that eat dead bodies. The bugs flock to the plant when it's in bloom, smelling its weird odor and hoping to eat.

Carrion flies are the corpse lily's pollinators. The flower needs them to pick up its pollen and carry it to another corpse lily. Even the flower's pollen is strange. **IT'S MORE LIKE STICKY SNOT THAN POWDER!**

BIG, STINKY, AND UNWELCOME

Not only is the corpse lily smelly: It's a parasitic plant! Corpse lilies are extreme parasites that don't have roots, stems, or leaves. They become so dependent on their host plant that they don't use photosynthesis to make food. Instead they weave threadlike body parts into hosts to steal nutrients.

CORPSE LILY SEEDS

RAFFLESIA ARNOLDII ALSO HAPPENS TO BE THE LARGEST FLOWER ON THE PLANET. IT CAN WEIGH UP TO 24 POUNDS (11 KG)! IT CAN RAISE ITS TEMPERATURE, TOO, PERHAPS CONVINCING BUGS EVEN MORE THAT IT'S ROTTING MEAT.

CAN'T TOUCH THIS!

Plants have adapted to harsh environments, tough competition, and limited resources, but you might not think humans have anything to fear from them. That's not true! There are plants that have adapted to be so dangerous that we're in trouble if we even brush past them.

A good example of a dangerous plant is the gympie gympie tree, common in parts of Australia. **IT'S ONE OF THE WORLD'S MOST TOXIC PLANTS.** It's covered with stinging hairs that can cause unbearable pain for months. Some have said it's the worst pain they've ever felt! Who knew plants could be so dangerous?

GYMPIE GYMPIE BERRIES

THE GYMPIE GYMPIE IS A SHRUB, BUT OTHER STINGING PLANTS ARE LARGE, LIKE THIS GIANT STINGING TREE!

CHAMPIONS OF CHANGE

Plants have found ways to adapt to many challenges in their habitats. Perhaps their greatest challenge is surviving **climate change**. Climate change affects temperatures and rainfall, and it has other consequences. You can help plants survive by finding ways to fight climate change.

GLOSSARY

climate change: long-term change in Earth's climate, caused partly by human activities such as burning oil and natural gas

develop: to cause something to grow or become bigger or more advanced

digestive: having to do with the body parts concerned with eating, breaking down, and taking in food

drought: a long period of time during which there is very little or no rain

environment: the natural world in which a plant or animal lives

fertilizer: something that makes soil better for growing crops and other plants

follicle: a dry fruit that opens on one side to release its seeds

fungus: a living thing that is somewhat like a plant, but doesn't make its own food, have leaves, or have a green color. Fungi include molds and mushrooms.

mineral: matter important in small amounts for the nutrition of plants and animals

nutrient: something a living thing needs to grow and stay alive

pollinate: to take pollen from one flower, plant, or tree to another

release: to set something free

tissue: matter that forms the parts of living things

FOR MORE INFORMATION

BOOKS

Aaseng, Nathan. *Weird Meat-Eating Plants*. Berkeley Heights, NJ: Enslow Publishers, 2011.

Rice, Barry. *Monster Plants: Meat Eaters, Real Stinkers, and Other Leafy Oddities*. New York, NY: Scholastic, 2010.

Whitehouse, Patricia. *Plants*. Chicago, IL: Heinemann Library, 2008.

WEBSITES

Easy Plants for Kids to Grow
www.hgtv.com/outdoors/flowers-and-plants/easy-plants-for-kids-to-grow-pictures
Here's a website with tips on how to grow your own plants!

Plants for Kids
www.sciencekids.co.nz/plants.html
This website has fun games, puzzles, and quizzes to test your plant knowledge.

INDEX

alga 7

anther 25

Australian grass tree 19

bacteria 7

banksia 18

bird orchid 24

botany 5

cactus 10, 11

carnivorous plants 20

chlorophyll 8

climate change 29

corpse lily 26, 27

Darwin, Charles 6

desert 10

desert rhubarb 13

environment 4, 6

epiphytes 8

false rose of Jericho 13

fungi 14

giant sequoia tree 16, 17

gympie gympie tree 28

mimicry 24

mistletoe 23

nectar 20

parasitic plants 22

photosynthesis 8, 9

phloem 8

pistil 25

pitcher plant 20

spines 11

Venus flytrap 20

xylem 8